# Protecting the Electric Grid from the Potential Threats of Solar Storms and Electromagnetic Pulse

I0482518

**Homeland Security and Governmental Affairs Committee**

Full Committee Hearing

**July 22, 2015 10:00AM**

Location: SD-342, Dirksen Senate Office Building

ISBN-10    1533220409

ISBN-13    978-1533220400

.

# Table of Contents

## Member Statements

## Witnesses

**Opening Statement of Chairman Ron Johnson**
**"Protecting the Electric Grid from the Potential Threats of Solar Weather and Electromagnetic Pulse"**
**July 22, 2015**

*As prepared for delivery:*

Good morning.  Thank you all for joining us today.  We will be looking at an issue that I believe is vital to national security—the extent to which the electric grid may be vulnerable to the threats of solar weather or a high-altitude electromagnetic pulse.

When it comes to critical infrastructure, there are several key sectors, often called the "lifelines," that essentially undergird and support all other sectors.  The energy sector is one of these crucial lifelines.  Without it, the other sectors would cease to function.  Our economy, our livelihoods and our ability to defend ourselves would be crushed.

Protecting the electric grid is a monumental challenge, and the threats facing it are many and varied.  The grid's physical infrastructure is necessarily spread throughout the nation and often cannot be protected from severe weather, sabotage or vandalism.  Likewise, utilities themselves are encountering an enormous task in protecting their computer networks and fighting off cyberattacks.

We also know that the potential consequence of any attack or event on the grid is very high.  In a real-life example, in 2003, a cascading failure across the grid in the Northeast left almost 50 million people without power, many for days.  One federal study identified nine critical substations that could be disabled and potentially bring down the entire U.S. grid for more than 18 months.

The threats of solar weather and high-altitude electromagnetic pulse are unique in that they can affect a vast region of the country. They may damage assets on the grid that are expensive, difficult and time-consuming to replace.

It is my goal that this hearing enable us to define the problem––that is, to identify how significant these threats are to our electric grid and our nation.  We need to understand how ready our nation is for these threats, and we need to evaluate potential opportunities to mitigate them.

Several reports over the last decade have highlighted just how bad these electromagnetic threats could be.  While we want to avoid fear-mongering, we don't want to take these issues lightly.

One study estimated severe solar weather could leave as many as 130 million people without power for years.  Similarly, the EMP Commission estimated that 90% of the U.S. population could die as a result of the consequences of a high-altitude electromagnetic pulse.  The electromagnetic pulse of a nuclear blast 300 miles above the U.S. could potentially reach the entire country.

While these numbers may be worst-case projections, we need to be sure we are adequately studying these threats and prioritizing them against others. It is not enough to hope they never occur.

There are opportunities for protecting the electric grid from these threats, but they are costly. The EMP Commission, for example, projected that hardening the grid could cost $2 billion. Compared to the likely economic impact of one of these events, these costs may well be worth it.

That being said, this hearing is not one in which we are exploring ways we can place stronger regulations on industry. After 31 years in manufacturing before I came to the Senate, I understand that the level of regulation on businesses already is burdensome and has serious negative unintended consequences. I hope to see industry and government working together to meet the common challenges facing critical infrastructure.

Our witness panel is well equipped to handle these questions today, bringing vast experience and study of these issues to bear. Thank you all for joining us, and I look forward to your testimony.

# Statement of Ranking Member Thomas R. Carper
*"Protecting the Electric Grid from the Potential Threats of Solar Storms and Electromagnetic Pulse"*
July 22, 2015

*As prepared for delivery:*

Threats to the homeland have evolved considerably over the past 15 years. In the months after 9/11, the most pressing threat to the homeland came from Al-Qaeda terrorists planning attacks from remote caves in Afghanistan. Today, the terror threat has become far more diverse.

Some terror groups are still seeking sophisticated attacks against high profile targets. Other groups, such as ISIS, are attempting to inspire extremists all over the world – including here in the United States – to carry out simple attacks within their own communities.

We are also being attacked daily in cyberspace. In many ways, we are dealing with an epidemic of online theft and fraud. This epidemic is growing at an alarming rate, as attacks become more sophisticated and disruptive.

And the challenges we faced with the recent global Ebola outbreak and our ongoing efforts to counter the spread of avian influenza remind us that threats to the homeland aren't just man-made. To address these evolving threats, we must always look to stay at least one step ahead of the bad guys, or in some cases, Mother Nature.

At the same time, we have to reluctantly accept the reality that our nation cannot protect against every threat, or potential threat, out there. Though we should always strive for perfection, we simply do not have the resources to achieve 100 percent security all of the time. That is why it is so critical that we prioritize our homeland defenses. We must focus on those threats that our experience and intelligence tell us are most likely to occur, and would have the gravest impact if, God forbid, they became a reality.

Today's hearing gives us an opportunity to assess two different potential threats to our electrical grid—man-made electromagnetic pulses, or 'EMPs,' and geomagnetic disturbances caused by space weather.

Each of these threats poses some degree of risk to our communities – that much is clear. Our job, however, is to assess that risk and figure out where these threats rank in the spectrum of everything else our country faces. For example, we must determine how likely electro and geomagnetic threats are to occur given our existing preparations and deterrents. And if they were to occur, how they could impact the homeland?

Answers to these basic questions become all the more important and urgent amid the horrific reminders of the existing challenges we face from domestic terrorism and homegrown violent extremism in our own communities —attacks like those that occurred recently in Chattanooga and Charleston.

I hope today we can make some progress on this front, and that our witnesses can provide us with a clear-eyed assessment of these threats. I look forward to an informative hearing.

###

# Heading Toward An EMP Catastrophe*

Ambassador R. James Woolsey
Chairman, Foundation for Defense of Democracies
Former Director of Central Intelligence

## Senate Homeland Security and Governmental Affairs Committee

Washington, DC
July 22, 2015

For over a decade now, since the Congressional EMP Commission delivered its first report to Congress eleven years ago in July of 2004, various Senate and House committees have heard from numerous scientific and strategic experts the consensus view that natural and manmade electromagnetic pulse (EMP) is an existential threat to the survival of the American people, that EMP is a clear and present danger, and that something must be done to protect the electric grid and other life sustaining critical infrastructures--immediately.

Yet this counsel and the cost-effective solutions proposed to the looming EMP threat have been ignored. Continued inaction by Washington will make inevitable a natural or manmade EMP catastrophe that, as the Congressional EMP Commission warned, could kill up to 90 percent of the national population through starvation, disease, and societal collapse.

Indeed, some actions taken by the Congress, the White House, and the federal bureaucracy are impeding solutions, making the nation more vulnerable, and helping the arrival of an EMP catastrophe. More about that later.

Why has Washington failed to act against the EMP threat? A big part of the problem is that policymakers and the public still fail to understand that EMP, and the catastrophic consequences of an EMP event, are not science fiction.

The EMP threat is as real as the Sun and as inevitable as a solar flare.

The EMP threat is as real as nuclear threats from Russia, China, North Korea, and Iran. Nuclear EMP attack is part of the military doctrines, plans and exercises of all of these nations for a revolutionary new way of warfare that focuses on attacking electric grids and civilian critical infrastructures--what they call Total Information Warfare or No Contact Wars, and what some western analysts call Cybergeddon or Blackout Wars.

The nuclear EMP threat is as real as North Korea's KSM-3 satellite, that regularly orbits over the U.S. on the optimum trajectory and altitude to evade our National Missile Defenses and, if the KSM-3 were a nuclear warhead, to place an EMP field over all 48 contiguous United States.

The EMP threat is as real as non-nuclear radiofrequency weapons that have already been used by terrorists and criminals in Europe and Asia, and no doubt will sooner or later be used here against America.

### A Clear And Present Danger

EMP, while still inadequately understood by policymakers and the general public, has been the subject of numerous major scientific and strategic studies. All of these warn by consensus that a natural or nuclear EMP, in the words of the Congressional EMP Commission, "Is one of a small number of threats that has the potential to hold our society seriously at risk" and "Is capable of causing catastrophe for the nation." Such is the warning not only of the Congressional EMP Commission, but of studies by the Congressional Strategic Posture Commission, the National Academy of Sciences, the Department of Energy, the National Intelligence Council, a U.S. Federal Energy Regulatory Commission report coordinated with the Department of Defense and Oak Ridge National Laboratory, and numerous other reports.

Yet a recent Wall Street Journal article (May 1, 2015) on NORAD moving back into Cheyenne Mountain and spending $700 million to further harden the mountain against a nuclear EMP attack from North Korea, received hundreds of comments from shocked readers, half of whom still think that EMP is science fiction.

***Nuclear EMP***.  We know that EMP is not science fiction but an existential threat that would have catastrophic consequences for our society because of high-altitude nuclear tests by the U.S. and Russia during the early Cold War, decades of underground nuclear testing, and over 50 years of tests using EMP simulators.  For example, in 1961 and 1962, the USSR conducted several EMP tests in Kazakhstan above its own territory, deliberately destroying the electric grid and other critical infrastructures over an area larger than Western Europe.  The Congressional EMP Commission based its threat assessment partially on using EMP simulators to test modern electronics--which the Commission found are over one million times more vulnerable than the electronics of the 1960s.

One prominent myth is that a sophisticated, high-yield, thermonuclear weapon is needed to make a nuclear EMP attack.  In fact, the Congressional EMP Commission found that virtually any nuclear weapon--even a primitive, low-yield atomic bomb such as terrorists might build--would suffice.  The U.S. electric grid and other civilian critical infrastructures--for example, communications, transportation, banking and finance, food and water--have never been hardened to survive EMP.  The nation has 18 critical infrastructures--all 17 others depend upon the electric grid.

Another big myth is that a sophisticated long-range missile is needed to deliver an EMP attack.  The iconic EMP attack detonates a single warhead about 300 kilometers high over the center of the U.S., generating an EMP field over all 48 contiguous United States.

However, any warhead detonated 30 kilometers high anywhere over the eastern half of the U.S. would collapse the Eastern Grid.  The Eastern Grid generates 75 percent of U.S. electricity and supports most of the national population.  Such an attack could be made by a short-range Scud missile launched off a freighter, by a jet fighter or small private jet doing a zoom climb, or even by a meteorological balloon.

According to a February 2015 article by President Ronald Reagan's national security brain trust--Dr. William Graham who was Reagan's Science Advisor and ran NASA, Ambassador Henry Cooper who was Director of the Strategic Defense Initiative, and Fritz Ermarth who was Chairman of the National Intelligence Council--North Korea and Iran have both practiced the iconic nuclear EMP attack against the United States.  Both nations have orbited satellites on south polar trajectories that evade U.S. early warning radars and National Missile Defenses. North Korea and Iran have both orbited satellites at altitudes that, if the satellites were nuclear warheads, would place an EMP field over all 48 contiguous United States.

Dr. Graham and his colleagues in their article warn that Iran should already even be regarded as having nuclear weapons and missiles capable of making an EMP attack against the U.S., or against any nation on Earth.

North Korea and Iran have also apparently practiced making a nuclear EMP attack using a short-range missile launched off a freighter. Such an attack could be conducted anonymously to escape U.S. retaliation--thus defeating nuclear deterrence.

***Natural EMP.*** We know that natural EMP from the Sun is real. Coronal mass ejections traveling over one million miles per hour strike the Earth's magnetosphere, generating geomagnetic storms every year. Usually these geo-storms are confined to nations at high northern latitudes and are not powerful enough to have catastrophic consequences. In 1989, the Hydro-Quebec Storm blacked-out half of Canada for a day causing economic losses amounting to billions of dollars.

However, we are most concerned about the rare solar super-storm, like the 1921 Railroad Storm, which happened before American civilization became dependent for survival upon electricity and the electric grid. The National Academy of Sciences estimates that if the Railroad Storm were to recur today, there would be a nationwide blackout with recovery requiring 4-10 years, if recovery is possible at all.

The most powerful geomagnetic storm on record is the 1859 Carrington Event. Estimates are that Carrington was about 10 times more powerful than the 1921 Railroad Storm and 100 times more powerful than the 1989 Hydro-Quebec Storm. The Carrington Event was a worldwide phenomenon, causing forest fires from flaring telegraph lines, burning telegraph stations, and destroying the just laid intercontinental telegraph cable at the bottom of the Atlantic Ocean.

If a solar super-storm like the Carrington Event recurred today, it would collapse electric grids and life-sustaining critical infrastructures worldwide, putting at risk the lives of billions.

NASA in July 2014 reported that two years earlier, on July 23, 2012 , the Earth narrowly escaped another Carrington Event. A Carrington-class coronal mass ejection crossed the path of the Earth, missing our planet by just three days. NASA assesses that the resulting geomagnetic storm would have had catastrophic consequences worldwide.

We are overdue for recurrence of another Carrington Event. The NASA report estimates that likelihood of such a geomagnetic super-storm is 12 percent per decade. This virtually guarantees that Earth will experience a catastrophic geomagnetic super-storm within our lifetimes or that of our children.

***Radio-Frequency Weapons (RFWs).*** Just as nuclear and natural EMP are not science fiction, we also know that the EMP threat from non-nuclear weapons, commonly called Radio-Frequency Weapons, is real. Terrorists, criminals, and even disgruntled individuals have already made localized EMP attacks using RFWs in Europe and Asia. Probably sooner rather than later, the RFW threat will come to America.

RFWs typically are much less powerful than nuclear weapons and much more localized in their effects, usually having a range of one kilometer or less. Reportedly, according to the Wall Street Journal, a study by the U.S. Federal Energy Regulatory Commission warns that a terrorist attack that destroys just 9 key transformer substations could cause a nationwide blackout lasting 18 months.

RFWs offer significant advantages over guns and bombs for attacking the electric grid. The EMP field will cause widespread damage of electronics, so precision targeting is much less necessary. And unlike damage from guns and bombs, an attack by RFWs is much less conspicuous, and may even be misconstrued as an unusual accident arising from faulty components and systemic failure.

Some documented examples of successful attacks using Radio Frequency Weapons, and accidents involving electromagnetic transients, are described in the Department of Defense *Pocket Guide for Security Procedures and Protocols for Mitigating Radio Frequency Threats* (Technical Support Working Group, Directed Energy Technical Office, Dahlgren Naval Surface Warfare Center):

--"In the Netherlands, an individual disrupted a local bank's computer network because he was turned down for a loan. He constructed a Radio Frequency Weapon the size of a briefcase, which he learned how to build from the Internet. Bank officials did not even realize that they had been attacked or what had happened until long after the event."

--"In St. Petersburg, Russia, a criminal robbed a jewelry store by defeating the alarm system with a repetitive RF generator. Its manufacture was no more complicated than assembling a home microwave oven."

--"In Kzlyar, Dagestan, Russia, Chechen rebel commander Salman Raduyev disabled police radio communications using RF transmitters during a raid."

--"In Russia, Chechen rebels used a Radio Frequency Weapon to defeat a Russian security system and gain access to a controlled area."

-- "Radio Frequency Weapons were used in separate incidents against the U.S. Embassy in Moscow to falsely set off alarms and to induce a fire in a sensitive area."

--"March 21-26, 2001, there was a mass failure of keyless remote entry devices on thousands of vehicles in the Bremerton, Washington, area...The failures ended abruptly as federal investigators had nearly isolated the source. The Federal Communications Commission (FCC) concluded that a U.S. Navy presence in the area probably caused the incident, although the Navy disagreed."

--"In 1999, a Robinson R-44 news helicopter nearly crashed when it flew by a high frequency broadcast antenna."

 --"In the late 1980s, a large explosion occurred at a 36-inch diameter natural gas pipeline in the Netherlands. A SCADA system, located about one mile from the naval port of Den Helder, was affected by a naval radar. The RF energy from the radar caused the SCADA system to open and close a large gas flow-control valve at the radar scan frequency, resulting in pressure waves that traveled down the pipe and eventually caused the pipeline to explode."

--"In June 1999 in Bellingham, Washington, RF energy from a radar induced a SCADA malfunction that caused a gas pipeline to rupture and explode."

--"In 1967, the *USS Forrestal* was located at Yankee Station off Vietnam. An A4 Skyhawk launched a Zuni rocket across the deck. The subsequent fire took 13 hours to extinguish. 134 people died in the worst U.S. Navy accident since World War II. EMI [Electro-Magnetic Interference] was identified as the probable cause of the Zuni launch."

--North Korea used an Radio Frequency Weapon, purchased from Russia, to attack airliners and impose an "electromagnetic blockade" on air traffic to Seoul, South Korea's capitol. The repeated attacks by RFW also disrupted communications and the operation of automobiles in several South Korean cities in December 2010; March 9, 2011; and April-May 2012 as reported in "Massive GPS Jamming Attack By North Korea" (*GPSWORLD.COM*, May 8, 2012).

***All Hazards Strategy.*** The Congressional EMP Commission recommended an "all hazards" strategy to protect the nation by addressing the worst threat--nuclear EMP attack. Nuclear EMP is worse than natural EMP and the EMP from RFWs because it combines several threats in one. Nuclear EMP has a long-wavelength component like a geomagnetic super-storm, a short-wavelength component like Radio-Frequency Weapons, a mid-wavelength component like lightning--and is potentially more powerful and can do deeper damage than all three.

Thus, protecting the electric grid and other critical infrastructures from nuclear EMP attack will also protect against a Carrington Event and RFWs. Moreover, protecting against nuclear EMP will also protect the grid and other critical infrastructures from the worst over-voltages that may be generated by severe weather, physical sabotage, or cyber-attacks.

### EMP--The Ultimate Cyber Weapon

Ignorance of the military doctrines of potential adversaries and a failure of strategic imagination is setting America up for an EMP Pearl Harbor that could easily be avoided--if we would only heed that terrorist sabotage of electric grids and cyber-attacks are early warning indicators. In fact, in the military doctrines, planning, and exercises of Russia, China, North Korea and Iran, nuclear EMP attack is the ultimate weapon in an all-out cyber operation aimed at defeating nations by blacking-out their electric grids and other critical infrastructures.

For example, Russian General Vladimir Slipchenko in his military textbook *No Contact Wars* describes the combined use of cyber viruses and hacking, physical attacks, non-nuclear EMP weapons, and ultimately nuclear EMP attack against electric grids and critical infrastructures as a new way of warfare that is the greatest Revolution in Military Affairs (RMA) in history.  Like Nazi Germany's Blitzkrieg ("Lightning War") Strategy that coordinated airpower, armor, and mobile infantry to achieve strategic and technological surprise that nearly defeated the Allies in World War II, the New Blitzkrieg is, literally and figuratively an electronic "Lightning War" so potentially decisive in its effects that an entire civilization could be overthrown in hours.  According to Slipchenko, EMP and the new RMA renders obsolete modern armies, navies and air forces.  For the first time in history, small nations or even non-state actors can humble the most advanced nations on Earth.

China's military doctrine sounds an identical theme.  According to People's Liberation Army textbook *World War, the Third World War--Total Information Warfare*, written by Shen Weiguang (allegedly the inventor of Information Warfare), "Therefore, China should focus on measures to counter computer viruses, nuclear electromagnetic pulse...and quickly achieve breakthroughs in those technologies...":

*With their massive destructiveness, long-range nuclear weapons have combined with highly sophisticated information technology and information warfare under nuclear deterrence....Information war and traditional war have one thing in common, namely that the country which possesses the critical weapons such as atomic bombs will have "first strike" and "second strike retaliation" capabilities....As soon as its computer networks come under attack and are destroyed, the country will slip into a state of paralysis and the lives of its people will ground to a halt.  Therefore, China should focus on measures to counter computer viruses, nuclear electromagnetic pulse...and quickly achieve breakthroughs in those technologies in order to equip China without delay with equivalent deterrence that will enable it to stand up to the military powers in the information age and neutralize and check the deterrence of Western powers, including the United States.*

Iran in a recently translated military textbook endorses the theories of Russian General Slipchenko and the potentially decisive effects of nuclear EMP attack some 20 times.  An Iranian political-military journal, in an article entitled "Electronics To Determine Fate Of Future Wars," states that the key to defeating the United States is EMP attack and that, "If the world's industrial countries fail to devise effective ways to defend themselves against dangerous electronic assaults, then they will disintegrate within a few years.":

*Advanced information technology equipment exists which has a very high degree of efficiency in warfare.  Among these we can refer to communication and information gathering satellites, pilotless planes, and the digital system....Once you confuse the enemy communication network you can also disrupt the work of the enemy command*

*and decision-making center. Even worse, today when you disable a country's military high command through disruption of communications you will, in effect, disrupt all the affairs of that country....If the world's industrial countries fail to devise effective ways to defend themselves against dangerous electronic assaults, then they will disintegrate within a few years... American soldiers would not be able to find food to eat nor would they be able to fire a single shot.* (Tehran, *Nashriyeh-e Siasi Nezami*, December 1998 - January 1999)

North Korea appears to have practiced the military doctrines described above against the United States--including by simulating a nuclear EMP attack against the U.S. mainland.  Following North Korea's third illegal nuclear test in February 2013, North Korean dictator Kim Jong-Un repeatedly threatened to make nuclear missile strikes against the U.S. and its allies.  In what was the worst ever nuclear crisis with North Korea, that lasted months, the U.S. responded by beefing-up National Missile Defenses and flying B-2 bombers in exercises just outside the Demilitarized Zone to deter North Korea.  On April 9, 2013, North Korea's KSM-3 satellite orbited over the U.S. from a south polar trajectory, that evades U.S. early warning radars and National Missile Defenses, at the near optimum altitude and location to place an EMP field over all 48 contiguous United States.  On April 16, 2013, the KSM-3 again orbited over the Washington, D.C.-New York City corridor where, if the satellite contained a nuclear warhead, it could project the peak EMP field over the U.S. political and economic capitals and collapse the Eastern Grid, which generates 75 percent of U.S. electricity.  On the same day, parties unknown used AK-47s to attack the Metcalf transformer substation that services San Francisco, the Silicon Valley, and is an important part of the Western Grid.  Blackout of the Western Grid, or of just San Francisco, would impede U.S. power projection capabilities against North Korea.  In July 2013, a North Korean freighter transited the Gulf of Mexico with two nuclear capable SA-2 missiles in its hold, mounted on their launchers hidden under bags of sugar, discovered only after the freighter tried to return to North Korea through the Panama Canal.  Although the missiles were not nuclear armed, they are designed to carry a 10 kiloton warhead, and could execute the EMP Commission's nightmare scenario of an anonymous EMP attack launched off a freighter.  All during this period, the U.S. electric grid and other critical infrastructures experienced various kinds of cyber-attacks, as they do every day and continuously.

North Korea appears to have been so bold as to use the nuclear crisis it deliberately initiated to practice against the United States an all-out cyber warfare operation, including computer bugs and hacking, physical sabotage, and nuclear EMP attack.

Just as Nazi Germany practiced the Blitzkrieg in exercises and during the Spanish Civil War (1936-1939), before surprising the Allies in World War II, so terrorists and state actors appear to be practicing now.  For example:

--On October 27, 2013, the Knights Templars, a criminal drug cartel, blacked-out Mexico's Michoacan state and its population of 420,000, so they could terrorize the people and paralyze the police.  The Knights, cloaked by the blackout, entered towns and villages and publicly executed leaders opposed to the drug trade.

--On June 9, 2014, Al Qaeda in the Arabian Peninsula used mortars and rockets to destroy transmission towers, plunging into darkness all of Yemen, a country of 16 cities and 24 million people.  It is the first time in history that terrorists put an entire nation into blackout, and an important U.S. ally, whose government was shortly afterwards overthrown by terrorists allied to Iran.

--In July 2014, according to press reports, a Russian cyber-bug called Dragonfly infected 1,000 electric power-plants in Western Europe and the United States for purposes unknown, possibly to plant logic bombs in power-plant computers to disrupt operations in the future.

--On January 25, 2015, terrorists blacked-out 80 percent of the electric grid in Pakistan, a nation of 185 million people, and a nuclear weapons state.

--On March 31, 2015, most of Turkey's 75 million people experienced a widespread and disruptive blackout, the NATO ally reportedly victimized by a cyber-attack from Iran.

On June 20, 2015, the New York Times reported that administration officials in a classified briefing to Congress on a cyber-attack from China, that stole sensitive U.S. Government data on millions of federal employees, was information warfare "on a scale we've never seen before from a traditional adversary." Yet this and the other ominous threats described above are already forgotten, or relegated to back page news, as policymakers and the public stumble on, seemingly shell-shocked and uncomprehending, to the latest cyber crisis.

We as a nation are not "connecting the dots" through a profound failure of strategic imagination.  Like the Allies before the Blitzkrieg of World War II, we are blind to the unprecedented existential threat that is about to befall our civilization--figuratively and literally, from the sky, like lightning.

### *Washington Dysfunction*

The Congressional EMP Commission recommended a plan to protect the national electric grid from nuclear EMP attack, that would also mitigate all lesser threats--including natural EMP, RFWs, cyber bugs and hacking, physical sabotage, and severe weather--for about $2 billion, which is what the U.S. gives away every year in foreign aid to Pakistan.  About $10-20 billion would protect all the critical infrastructures from nuclear EMP attack and other threats.

There are other plans that cost much less, and much more, because there are different technologies and strategies for protecting against EMP, and to different levels of risk.  Any or all of these plans are commendable.  There is no such thing as being over-prepared for an existential threat.

Unfortunately, none of these plans has been implemented.  The U.S. electric grid and other civilian critical infrastructures remain utterly vulnerable to EMP because of lobbying by the electric utilities in Congress, the federal bureaucracy, and the White House.

Lobbying by the electric power industry and their North American Electric Reliability Corporation (NERC) has, so far, thwarted every bill by the U.S. Congress to protect the grid from EMP. For example, in 2010, the House passed unanimously the GRID Act-- which was denied a vote in the Senate, because a single Senator on the Energy and Natural Resources Committee put a hold on the bill. If the GRID Act passed in 2010, the national electric grid would already be protected from EMP, a process the EMP Commission estimated would take about 3-5 years.

The SHIELD Act, another bipartisan bill to protect the electric grid, has been stalled in the House Energy and Commerce Committee for years, due to lobbying by the electric utilities.

Even worse, the U.S. Federal Energy Regulatory Commission, which has a too deferential and too cozy relationship with NERC, has approved a NERC proposed standard for protecting the grid from solar storms that has been condemned by the best scientific experts. Dr. William Radasky and John Kappenman, who both served on the Congressional EMP Commission, and other independent experts have written scientific critiques proving that the NERC standard for natural EMP (also called GMD for Geo-Magnetic Disturbance) is based on "junk science" that grossly underestimates the threat from natural EMP.

For example, Kappenman and Radasky, who are among the world's foremost scientific and technical experts on geomagnetic storms and grid vulnerability, warn that NERC's GMD Standard consistently underestimates the natural EMP threat from geo-storms: "When comparing...actual geo-electric fields with NERC model derived geo-electric fields, the comparisons show a systematic under-prediction in all cases of the geo-electric field by the NERC model."

Dr. Radasky, who holds the Lord Kelvin Medal for setting standards for protecting European electronics from natural and nuclear EMP, and John Kappenman, who helped design the ACE satellite upon which industry relies for early warning of geomagnetic storms, conclude that the NERC GMD Standard so badly underestimates the natural EMP threat that "its resulting directives are not valid and need to be corrected." Kappenman and Radasky:

*These enormous model errors also call into question many of the foundation findings of the NERC GMD draft standard. The flawed geo-electric field model was used to develop the peak geo-electric field levels of the Benchmark model proposed in the standard. Since this model understates the actual geo-electric field intensity for small storms by a factor of 2 to 5, it would also understate the maximum geo-electric field by similar or perhaps even larger levels. Therefore, the flaw is entirely integrated into the NERC Draft Standard and its resulting directives are not valid and need to be corrected.*

The excellent Kappenman-Radasky critique of the NERC GMD Standard represents the consensus view of all the independent observers who participated in the NERC GMD Task Force.

Perhaps most revelatory of U.S. FERC's failures, by approving the NERC GMD Standard that grossly underestimates the natural EMP threat from geo-storms--U.S. FERC abandoned its own much more realistic estimate of the natural EMP threat from geo-storms. It is incomprehensible why U.S. FERC would ignore the findings of its own excellent interagency study, one of the most in depth and meticulous studies of the EMP threat ever performed, that was coordinated with Oak Ridge National Laboratory, the Department of Defense, and the White House.

U.S. FERC's preference for NERC's "junk science" over U.S. FERC's own excellent scientific assessment of the geo-storm threat is indefensible.

The White House has not helped matters by issuing a draft executive order for protecting the national grid from natural EMP--but that trusts NERC and the electric utilities to set the standards.

Nor has the White House or the U.S. FERC challenged NERC's assertion that it has no responsibility to protect the electric grid from nuclear EMP or Radio-Frequency Weapons.

Nor has the White House or the U.S. FERC done anything to prevent NERC and the utilities from misinforming policymakers and the public about the EMP threat and their lack of preparedness to survive and recover from an EMP catastrophe.

Consequently, policymakers in the States who are alarmed by the lack of progress in Washington on EMP preparedness, find themselves seriously disadvantaged in efforts to protect their State electric grids by the utilities and their well-funded lobbyists who falsely claim Washington and the utilities are making great progress partnering on the EMP problem. So far in 2015, State initiatives to protect their electric grids have been defeated by industry lobbyists in Maine, Colorado, and Texas.

Texas State Senator Bob Hall, a former USAF Colonel and himself an EMP expert, characterizes as "equivalent to treason" the behavior of the electric utilities and their lobbyists:

*As a Texas State Senator who tried in the 2015 legislative session to get a bill passed to harden the Texas grid against an EMP attack or nature's GMD, I learned firsthand the strong control the electric power company lobby has on elected officials. We did manage to get a weak bill passed in the Senate but the power companies had it killed in the House. A very deceitful document which was carefully designed to mislead legislators was provided by the power company lobbyist to legislators at a critical moment in the process. The document was not just misleading, it actually contained false statements. The EMP/GMD threat is real and it is not "if" but WHEN it will happen. The responsibility for the catastrophic destruction and wide spread death of*

*Americans which will occur will be on the hands of the executives of the power companies because they know what needs to be done and are refusing to do it. In my opinion power company executives, by refusing to work with the legislature to protect the electrical grid infrastructure are committing an egregious act that is equivalent to treason. I know and understand what I am saying. As a young US Air Force Captain, with a degree in electrical engineering from The Citadel, I was the project officer who led the Air Force/contractor team which designed, developed and installed the modification to "harden" the Minuteman Strategic missile to protect it from an EMP attack. The American people must demand that the power company executives that are hiding the truth stop deceiving the people and immediately begin protecting our electrical grid so that life as we know it today will not end when the terrorist EMP attack comes.*

Ironically, while electric power lobbyists are fighting against EMP protection in Washington, Texas, Maine, Colorado and elsewhere, the Iranian news agency MEHR recently reported that Iran is violating international sanctions and going full bore to protect itself from a nuclear EMP attack:

*Iranian researchers...have built an Electromagnetic Pulse (EMP) filter that protects country's vital organizations against cyber-attack. Director of Kosar Information and Communication Technology Institute Saeid Rahimi told MNA correspondent that the EMP (Electromagnetic Pulse) filter is one of the country's boycotted products and until now procuring it required considerable costs and various strategies. "But recently Kosar ICT...has managed to domestically manufacture the EMP filter for the very first time in this country," said Rahimi. Noting that the domestic EMP filter has been approved by security authorities, Rahimi added "the EMP filter protects sensitive devices and organizations against electromagnetic pulse and electromagnetic terrorism." He also said the domestic EMP filter has been implemented in a number of vital centers in Iran. (MEHR News Agency, "Iran Builds EMP Filter for 1st Time" June 13, 2015)*

### What Is To Be Done?

**Congress should pass the Critical Infrastructure Protection Act (CIPA),** which requires the Department of Homeland Security to adopt a new National Planning Scenario focused on EMP; to develop plans to protect the critical infrastructures; and for emergency managers and first responders to plan and train to protect and recover the nation from an EMP catastrophe. CIPA will enable DHS to draw upon the deep expertise within the Department of Defense and the Intelligence Community to help protect the critical infrastructures from EMP. Do not let the electric power lobby defeat CIPA or weaken its provisions, as they are presently trying to do.

**Reestablish the Congressional EMP Commission.** The greatest progress was being made when the EMP Commission existed to advance EMP preparedness. Progress stopped when the EMP Commission terminated in 2008. Currently, the struggle to advance national EMP preparedness is being carried on by a handful of patriotic individuals and Non-Government Organizations who have no official standing and extremely limited resources. Bring back the EMP Commission with its deep expertise to advise Congress, government at all levels, and the private sector on how best to protect the nation, and to serve as a watchdog and leader for national EMP preparedness.

**Pass the SHIELD Act or the GRID Act** to establish adequate regulatory authority within the U.S. Government to achieve timely protection of the electric grid--and watch U.S. FERC like a hawk to make sure that regulatory authority is exercised.

**Include in the National Defense Authorization Act the simple two-sentence provision below,** that could rapidly reverse the trend of America's increasing vulnerability to EMP, by directing the Secretary of Defense to help State governments and the electric utilities protect themselves from an EMP catastrophe:

*Energy Security For Military Bases And Critical Defense Industries*
*Whereas 99 percent of the electricity used by CONUS military bases is supplied by the national electric grid; whereas the Department of Defense has testified to Congress that DoD cannot project power overseas or perform its homeland security mission without electric power from the national grid; whereas the Congressional EMP Commission warned that up to 9 of 10 Americans could die from starvation and societal collapse from a nationwide blackout lasting one year; therefore the Secretary of Defense is directed to urge governors, state legislators, public utility commissions of the 50 states, the North American Electric Reliability Corporation (NERC) and the utilities that supply electricity to CONUS military bases and critical defense industries, to protect the electric grid from a high-altitude nuclear electromagnetic pulse (EMP) attack, from natural EMP generated by a solar super-storm and from other EMP threats including radiofrequency weapons, and to help the states, NERC, public utilities commissions, and electric utilities by providing DoD expertise on EMP and other such support and resources as may be necessary to protect the national electric grid from natural and manmade EMP threats. The Secretary of Defense is authorized to spend up to $2 billion in FY2017 to help protect the national electric grid from EMP.*

Ambassador R. James Woolsey is former Director of Central Intelligence and is Chairman of the Foundation for Defense of Democracies.

*I am highly indebted to my friend and colleague, Dr. Peter Vincent Pry, who served on the Congressional EMP Commission and is Executive Director of the EMP Task Force on National and Homeland Security, for assistance in drafting this testimony.

**Testimony of Joseph McClelland**
**Director, Office of Energy Infrastructure Security**
**Federal Energy Regulatory Commission**
**Before the Committee on Homeland Security**
**and Governmental Affairs**
**United States Senate**
**July 22, 2015**

Chairman Johnson, Ranking Member Carper and Members of the

Committee:

Thank you for the privilege to appear before you today to discuss threats to

the electric grid in the United States. My name is Joe McClelland and I am the

Director of the Federal Energy Regulatory Commission's newest office, the Office

of Energy Infrastructure Security. I am here today as a Commission staff witness

and my remarks do not necessarily represent the views of the Commission or any

individual Commissioner.

In the Energy Policy Act of 2005, Congress entrusted the Commission with

a major new responsibility to approve and enforce mandatory reliability standards

for the Nation's bulk power system. This authority is in section 215 of the Federal

Power Act. It is important to note that FERC's jurisdiction and reliability

authority under section 215 is limited to the "bulk power system," as defined in

the FPA, which excludes Alaska and Hawaii, as well as local distribution systems.

Under the section 215 authority, FERC cannot author or modify reliability

standards, but must depend upon an Electric Reliability Organization (or ERO) to perform this task. The Commission certified the North American Electric Reliability Corporation or NERC as the ERO. The ERO develops and proposes reliability standards or modifications for the Commission's review which it can either approve or remand. If the Commission approves a proposed reliability standard, it becomes mandatory in the United States and is applicable to the users, owners and operators of the bulk power system. If the Commission remands a proposed standard, it is sent back to the ERO for further consideration. The Commission is required to give "due weight" to the technical expertise of the ERO when reviewing any of NERC's proposed standards.

Section 215 of the Federal Power Act provides a statutory foundation for the ERO to develop reliability standards for the bulk power system. However, the nature of a national security threat by entities intent on attacking the U.S. by exploiting vulnerabilities in its electric grid using physical or cyber means stands in stark contrast to other major reliability events that have caused regional blackouts and reliability failures in the past, such as events caused by tree trimming practices. Widespread disruption of electric service can quickly undermine the U.S. government, its military, and the economy, as well as endanger the health and safety of millions of citizens. Given the national security dimension to this threat, there may be a need to act quickly to protect the grid in a manner where action is mandatory rather than voluntary while protecting certain sensitive information from public disclosure.

To provide a significantly more agile and focused approach to these growing cyber and physical security threats, the Commission established the Office of Energy Infrastructure Security – or OEIS – in late 2012. Its mission is to provide leadership, expertise and assistance to the Commission, other federal and state agencies and jurisdictional entities in identifying, communicating and seeking comprehensive solutions to significant potential cyber and physical security risks to the energy infrastructure under the Commission's jurisdiction. This includes threats from geomagnetic disturbances (GMDs) and electromagnetic pulses (EMPs). OEIS also assists in the identification of key energy infrastructure facilities for the application of best practices. OEIS has been able to recruit and develop deep subject matter expertise to collaboratively perform its task.

Specific to the subject of this hearing, GMD and EMP events are generated from either naturally occurring or man-made causes. In the case of GMDs, naturally occurring solar magnetic disturbances periodically disrupt the earth's magnetic field which in turn, can induce currents on the electric grid that may simultaneously damage or destroy key transformers over a large geographic area. Regarding man-made events, EMPs can be generated by devices that range from small, portable, easily concealed battery-powered units all the way through missiles equipped with nuclear warheads. In the case of the former, equipment is readily available that can generate localized high-energy bursts designed to disrupt, damage or destroy electronics such as those found in control systems on the electric grid. The EMP generated during the detonation of a nuclear device is far more encompassing and generates three distinct effects, each impacting different types of equipment; a short high energy RF-type burst called E1 that destroys electronics; a slightly longer burst that is similar to lightning termed E2; and a final effect termed E3 that is similar in character and effect to GMD targeting the same equipment including key transformers. Any of these effects can cause voltage problems and instability on the electric grid, which can lead to wide-area blackouts.

In 2001, Congress established a commission to assess and report on the threat from EMP. In 2004 and again in 2008, the commission issued reports on these threats. One of the key findings in the reports was that a single EMP attack could seriously degrade or shut down a large part of the electric power grid. Depending upon the attack, significant parts of the electric infrastructure could be "out of service for periods measured in months to a year or more." It is important to note that effective mitigation against solar geomagnetic disturbances and non-nuclear EMP weaponry can also provide an effective mitigation against the impacts of a high-altitude nuclear detonation.

In order to better understand and quantify the effect of EMP and GMD on the power grid, FERC staff, the Department of Energy and the Department of Homeland Security sponsored a study conducted by the Oak Ridge National Laboratory in 2010. The results of the study support the general conclusion of prior studies that EMP and GMD events pose substantial risk to equipment and operation of the Nation's electric grid and under extreme conditions could result in major long-term electrical outages. Unlike EMP attacks that are dependent upon the capability and intent of an attacker, GMD disturbances are inevitable with only the timing and magnitude subject to variability. The Oak Ridge study assessed a solar storm that occurred in May 1921, which has been termed a 1-in-100 year event, and applied it to today's electric grid. The study concluded that such a storm could damage or destroy over 300 bulk power system transformers interrupting service to 130 million people with some outages lasting for a period of years.

The Commission has used a two-fold approach to help address the EMP and GMD threats:

1. In response to a Commission Order, NERC has proposed two reliability standards on GMD. The Commission approved the first one, a mandatory reliability standard that requires certain entities to implement operational procedures to mitigate the effects of GMD events. The Commission also has issued an order proposing to approve the second one, a reliability standard proposed by NERC that would establish requirements for certain entities to conduct initial and on-going assessments of the vulnerability of their transmission systems against a

benchmark geomagnetic disturbance. The Commission also proposed certain additional actions.

2. Simultaneous with its regulatory approach, the Commission collaborated with federal agencies and industry members to identify key energy facilities, conduct threat briefings to industry members on both GMD and EMP, assist with the identification of best practices for mitigation, and cooperate with international partners to convey threat and mitigation information as well as encourage adoption of best practices for mitigation.

A few US entities have taken some initial steps to address EMP on their systems, but much work remains. Internationally, the United Kingdom, Norway, Sweden, Finland, Germany, South Korea, Japan, Australia, New Zealand, South Africa, Israel and Saudi Arabia have GMD and/or EMP programs in place or are in the early stages of addressing or examining the impacts of GMD or EMP.

The costs of these initiatives can vary widely depending on factors such as the threshold of protection, the service requirements of the load, the type of equipment that is to be protected, and whether the installation is new or a retrofit.

In conclusion, these types of threats pose a serious risk to the electric grid and its supporting infrastructures that serve our Nation. The Commission is therefore taking both regulatory and collaborative actions to address them.

Thank you again for the opportunity to testify today.  I would be happy to answer any questions you may have.

Richard L. Garwin
IBM Fellow Emeritus[1]
IBM Thomas J. Watson Research Center
P.O. Box 218, Yorktown Heights, NY  10598
www.fas.org/RLG/
Email: RLG2@us.ibm.com

Prepared testimony for the hearing,
"Protecting the Electric Grid from the Potential Threats of Solar Storms and Electromagnetic Pulse"

The spectacular images of Pluto this week from the NASA New Horizons probe provoked great public interest in our solar system.  But our solar system is a matter for concern, as well. The 1200 people injured February 15, 2013 at Chelyabinsk, Russia, from a bolide (meteor) brought substantial focus on low-probability, high-consequence events. Among these are particularly intense magnetic storms from space-weather events or coronal mass ejections (CME), possibly even more intense than the 1859 Carrington Event in the pre-electric-grid era

Another potentially great impact on the electrical grid and modern societies is the high-altitude electromagnetic pulse (HEMP) from high-altitude nuclear explosions—HANE—on the order of 100 km or more above the Earth's surface.

The United States has been a leader in long-distance transmission of electrical power, but its system differs in characteristics, management, and organization from those of other advanced states.  Nevertheless, there is much to be learned from and by the United States in working to make our electrical grid robust and economical in the modern era of technological threats and opportunities.

I begin with my recommendations to ease and essentially solve the severe problem posed by geomagnetic storms induced by space weather—specifically by the routine ejection from the sun of enormous blocks of plasma that travel out within the solar system and reach the Earth typically in a couple of days[2].  These cause displays of the "Northern Lights" (and Southern Lights as well).  More importantly, the magnetized plasma and its incorporated magnetic field merge with the magnetic field of the Earth and change it by a relatively small amount, which, however, can create large currents on long electrical conductors such as pipelines, telegraph wires in the old days, and the electrical power transmission system—the Bulk Power System.

[1] Affiliation given for identification only.
[2] See *"Impacts of Severe Space Weather on the Electric Grid,"* JSR-11-320 of November 2011, sponsored by DHS, of which I was an author—available at https://fas.org/irp/agency/dod/jason/spaceweather.pdf  A broad set of recommendations may be viewed on pp. 3-5 of that report.

Very serious consequences are estimated for such an event of a magnitude that can be expected to occur at random once per century, with greater events occurring with lower probability and lesser events more frequently[3].

I emphasize that a "once per century" event might occur next week; it has a probability of 10% of occurring within the next ten years—a time in which we can and should take measures to reduce and essentially eliminate its impact on the Bulk Power System of the United States. But events expected to occur once in 20 years can cause significant damage and disruption.

**My recommendations regarding the Bulk Power System[4].**

*Missing in Federal policy and practice is a program to*

1. *train and equip utility and transmission operators to bring down within seconds (switch off) transmission lines that are at risk of being damaged.*

2. *implement "rapid islanding" of the grid, to maintain a large fraction of the power consumers in operation by the use of whatever island generation capacity exists; this also facilitates restoring the Bulk Power System to operation, in contrast with a "black start."*

3. *fit transmission lines on a priority basis with "neutral current blocking devices" (capacitors) in the common neutral-to-ground link of the 3-phase transformers of EHV transmission systems at one end of the line-- whether 3-phase transformers or 3 single-phase transformers. Where transformers at both ends are autotransformers this may not be possible, in which case series-blocking capacitors in the power lines themselves should be installed (even if shorted until an EMP event is recognized).*

4. *alert grid operators and others to a high-altitude nuclear explosion within milliseconds of the event (by detection of the unambiguous very brief E1—pronounced "Ee-one"--pulse).*

In my supplemental testimony submitted for the record, I provide support for these recommendations and explain why they would largely and immediately also eliminate long-lasting damage to the EHV transmisions system that might otherwise result from a high-altitude nuclear explosion.

** End of prepared oral testimony **

---

[3] It is important to understand what can and can not be done to mitigate damage from events that we wish would never happen, as was done in exemplary fashion in the FEMA-sponsored publication *"Key Planning Factors: Response to an Improvised Nuclear Device [explosion] in the National Capital Region"* November 2011, http://www.fas.org/irp/agency/dhs/fema/ncr.pdf

[4] I note that these recommendations are similar to those of the "E-PRO HANDBOOK" Executive Summary 2014 and the INTERNATIONAL E-PRO REPORT of September 2013, e.g.,
> *GIC current blockers*
> *Series Capacitance*
> *Reducing Transformer Loads*
> *Real-time, Threshold-based Transformer Protection*

Permanent and severe damage to the Bulk Power System occurs largely from the destruction of the extremely high voltage—EHV—transformers that are used to transmit the high-voltage alternating current three-phase power over distances of hundreds of miles. The electricity in our houses, offices, and factories is delivered from the wall plug at a voltage of 120 or 240 V, and large motors, trains, and other system generally consume electrical power at a voltage of some hundreds of volts. But because power is voltage multiplied by current—specifically watts equal volts-times-amperes, and megawatts equals kilovolts-times-kilo-amperes, the only way to transmit electrical power economically over a distance of 100 miles or more is to use a *transformer* to step up the voltage from the convenient generating level of a few thousand volts—kilovolts or kV—to EHV levels exceeding 500 kV.

The Earth's magnetic field changes irregularly over a period of minutes and hours and even days in the course of a geomagnetic storm, and by Faraday's law of magnetic induction produces small voltages in potential electrical circuits—voltages that are totally imperceptible to people and that in our automobiles, homes, or offices are of no concern. But according to Faraday (and this is the principle upon which all electrical motors and transformers are made) the voltage induced is proportional not only to the change of magnetic field per second of time, but to the area of the electrical circuit (and to the number of "turns" of wire around that circuit).

In the case of long-distance power lines that may be 50 meters (164 ft)—above ground, there is a substantial area of the circuit that might be expected to be the height of the power line above the ground, multiplied by the length of the transmission line in hundreds of kilometers. in fact, the area is far greater because, for these slow changes of magnetic field, the voltage around the closed circuit that is composed of the power lines on the transmission towers, and completed by the return of electrical current through the "ground," does not flow along the surface of the Earth. Rather it flows along the higher conductivity regions that are found at depths of 100-200 km or more in regions of the continents overlain by highly insulating crystal and rock such as granite. Much of the geology of eastern Canada and the northeast United States is of this nature, and so the "circuit" area for the changing magnetic field to do its dirty work may be 1000 km long by 100 km high--the size of a small state tipped on its side; the area is not 1000 km by 50m but 2,000 times as large!

The resulting voltage around the one-turn circuit is often expressed as the length of the line multiplied by the "electric field" expressed in volts per kilometer—V/km, and a geo-electric field as small as 5 V/km can cause serious damage because over a line of length 1000 km it would amount to 5,000 V. The particular vulnerability of transformers on the Bulk Power System arises when they are connected on the three-phase line so that the three fat aluminum power cables at the top of the poles enter three separate transformers that are "Y-connected," with their common point connected to a grounding mat or a field of metal stakes driven into the ground. Two such sets of Y-connected transformers at either end of the 1000-km line thus establish a circuit for the geomagnetic storm to induce current.

Despite the fact that EHV transmission occurs at voltages of 500 kV, and we have estimated 5 kV for the voltage due to the geomagnetic storm, the geomagnetic storm voltage is akin to "direct current" like that from a battery, whereas the power carried by the EHV system is alternating current, changing direction (twice) 60 times per second—at 60 hertz (Hz).

Over a period of many seconds or minutes, the dc current drives the transformers into "half-cycle saturation" allowing unprecedented amounts of power to flow from the generators or the source of electrical power, and overheating the copper windings and steel structure of the transformers.

It is essential to understand that geomagnetic storms cause no problems when the transmission systems are de-energized, as they would be following the downing of a transmission tower. Hence the first recommendation.

The second is to avoid collapse of the entire economy—blackout due to the loss of the most vulnerable line from the effect of geomagnetic storm, HANE, sabotage, or other problem. There is a big difference in the recovery time of the electrical power system between the blackout of an area covering many states and eastern Canada, and the loss of EHV transmission lines that only supplement more local generation capacity.

In fact, all but the most intense geomagnetic storm can be countered and Bulk Power Transmission continued if the Y-connected transformers are not connected from their common "neutral" terminal directly to the grounding mat, but instead through a "neutral current blocking device" that is designed to accept for a few minutes or hours steady voltages that could be expected from the 100-year geomagnetic storm. There have been several successful trials of such blocking devices in the United States, Canada, and elsewhere, and they are now offered for sale to the industry.

Their cost is on the order of $100,000 per tranformer[5], but they protect transformers that at a high-power terminal may cost $10 million[6] and can preserve the economy of a million Americans that would otherwise suffer from temporary disruption if the power line needed to be shut off, and severe economic loss and even loss of employment and life if the geomagnetic storm or HANE is allowed to destroy many transformers that would take months or years to replace.

Finally, essentially all transformer damage from a high-altitude nuclear explosion could be avoided by the installation of these blocking devices, or even where no such devices were installed, by manual or automatic shutdown of that EHV line for a minute or so following the detection of a HANE.

---

[5] http://www.powerworld.com/files/06Emprimus.pdf
[6] http://energy.gov/sites/prod/files/Large%20Power%20Transformer%20Study%20-%20June%202012_0.pdf A single-phase 500 MW large power transformer is quoted at $4.5 million.

**Protection of U.S. society against a high-altitude nuclear explosion.**

Such a high-altitude nuclear explosion—HANE—provides disturbances to long-distance power transmission systems by virtue of the high-altitude electromagnetic pulse—HEMP—through mechanisms that are complex and fascinating, but can be understood in broad outline and that have been the subject of much analysis over the decades since they were observed in fragmentary form in 1962.

A nuclear weapon exploded 100 km or so above the surface of the United States or above its shores would have "line of sight" out to 1000 km or so. This applies to a "normal" first-generation nuclear weapon as well as to a megaton-class nuclear explosive such as possessed by the United States, China, and Russia.

The geomagnetic-storm-like effect of a HANE arises from the liberation of large amounts of energy in a small (say one ton) mass of bomb and rocket materials in the weak magnetic field of the Earth. A magnetic bubble, 100-km or more in diameter, forms and is squeezed by the diverging magnetic field—the motion of these field lines in some sense mimics the disturbance formed by the incorporation of a portion of the magnetized plasma from the coronal mass ejection into the Earth's magnetosphere. The details of the resulting magnetic and electric disturbances on Earth are exquisitely complex because the bomb itself, before the expansion can take place, has liberated most of its energy in the form of vast amounts of soft x-rays that increase the ionization at the top of the atmosphere and serve largely to shield against the magnetic field variation from the "bubble" and "heave" of the bomb plasma in the magnetic field of the Earth. The resulting slow component of the electrical field from a HANE is dubbed E3. The time scale is typically ten seconds or more.

As might be suspected, there is an E1, which comes from the prompt gamma rays from the fission process. Within less than a nanosecond of an individual fission, a couple of percent of the energy release is emitted as the equivalent of extremely high voltage x-rays such as those used for radiography and radiotherapy. In a nuclear explosive—warhead or bomb—most of the gamma rays are absorbed, but those high-energy gamma rays that do emerge travel radially from the explosion above the atmosphere, although more might travel up or down or sideways depending upon the detailed internal design of the bomb. The bulk of the gamma rays may emerge over a few-nanosecond interval.

In 1962 the effect of the resulting E1 was observed in Hawaii, 1000 km from the explosion in space of a 1.4 megaton hydrogen bomb at an altitude of 400 km.

In contrast to earlier predictions of a modest electromagnetic pulse from a space nuclear explosion, on the order of 1 V/m at 1000 km[7], the detected EMP in this very fast-time (high frequency) range was of the order of 5,000 V/m, which was unexplained for many months after it

---

[7] R.L. Garwin, "*Determination of Alpha by Electro-magnetic Means,*" Los Alamos Scientific Lab., Report LAMS-1871, (1954), S-RD.

had been observed, until Los Alamos physicist Conrad Longmire, in preparing for a talk at the Air Force weapons lab in Albuquerque, thought of the mechanism by which such efficient conversion of gamma ray energy to electromagnetic pulse could be achieved.

Although an observder anywhere on Earth within line of sight to the space explosion receives this radio pulse as if it came directly from the explosion itself, it really originates in the upper atmosphere on the line of sight from the bomb to the observer. As the gamma rays produce fast electrons from the molecules of air, the electrons travel initially along the line of sight outward from the bomb, but their paths are *curved* by the weak magnetic field of the Earth. These curved paths radiate, but it seems at first thought impossible that electrons materializing over a path length of 10 km (flight time of 30 microseconds) could add their signals in the nanosecond range, but that is exactly what happens--because the electrons travel at nearly the speed of light, and the gamma rays, which materialize all along this 10 km path, travel at the speed of light in vacuum so that the radio wave is strengthened until the gamma rays are extinguished by absorption in the air atoms.

The result can be the conversion of 10% of the gamma ray energy into electromagnetic pulse, and clever bomb designers can make this pulse even shorter than is natural for an ordinary fission bomb.

However, the EHV transmission system has no special vulnerability to this E1 fast pulse. It was thoroughly addressed and emphasized by the EMP Commission Report of 2008,

**Impact of *E1* on critical infrastructure**

No mechanism has been identified and there is no experimental or theoretical reason to judge that even the most intense *E1* field will cause direct harm to humans or animals. Furthermore, there is a much shielding of sensitive electronics to electric fields in this range. The EMP Commission arranged for experimental tests of exposure of various kinds of electronics to EMP simulators—specifically *E1*.

Of 37 gasoline-fueled automobiles, 3 stopped running when exposed to simulated *E1*, but all restarted without incident. No effects were observed on cars not running during the EMP exposure. Similar results were obtained for trucks.

With regard to the electrical grid, electromagnetic relays that sense current and voltage by means of the forces produced by their magnetic fields, were immune to $E_1$. About the more modern electronic relays, the full unclassified 2008 EMPC report, "Critical National Infrastructures" states (p. 40):

> *"Electronic protective relays.* These devices (see figure 2-5) are the essential elements preserving high-value transmission equipment from damage during geomagnetic storms and other modes of grid collapse. Fortunately, these test items were the most robust of any of the electronic devices tested. However, test agencies reported that they are subject

to upset at higher levels of simulated EMP exposure. We believe that altering the deployment configurations can further ameliorate the residual problems."

Thus, relatively simple field retrofits would preserve the electronic protective relays; however, the power grid is imperiled by unnecessarily weak links.

Consumer electronics in operation will suffer upset or damage at E1 fields of some 10kV/m. The EMPC report cites the RS-232 ports of PCs (personal computers) as particularly vulnerable, and PCs are used in the SCADA (systems control and data acquisition) facilities of the electrical grid and other industries, so a robust Bulk Power System will require protective filters on the control computers.

Other nations have taken more seriously improving the resilience of their Bulk Power Systems against geomagnetic storms (and hence E3 from a high-altitude nuclear explosion), as detailed in (4). In this effort there are major technological opportunities to reduce cost of protection and prediction.

One of the substantial lacks in planning and operation to reduce space weather impact on the grid is adequate and continuous magnetic field data, as well as corresponding measurements of GIC. GIC measurements must be obtained from the power transmission companies, and that is in process, but particularly in the United States is bureaucratically difficult. On the other hand, magnetometer data has become easier and cheaper to obtain, as the result of the universal deployment of SmartPhones containing a compass, which is a three-component magnetometer. So here is a reference to and a trace in frequency of the background magnetic noise from anisotropic magneto resistance (AMR) sensor in a typical SmartPhone.

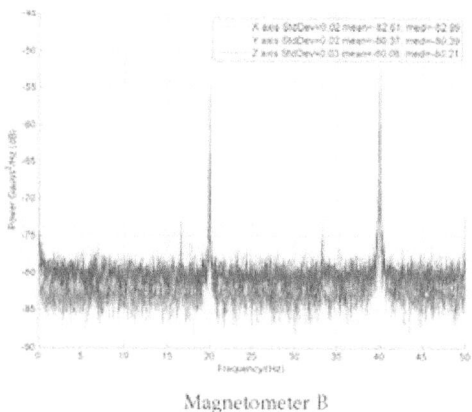

Magnetometer B

These SmartPhones can be programmed to record the magnetic field in an intelligent way, and to transmit it over the Web, either as a typical data call, or via WiFi in case the magnetometer is located close to some facility.

So rather than think of deploying classical magnetometers, one should include the possibility of the SmartPhone magnetometer produced by the millions and correspondingly cheap and robust.

Furthermore, some of the approaches to eliminating geomagnetic-storm-induced current (GIC) are not well appreciated—for instance the use of series capacitors in the three-phase power lines themselves, where blocking the path from transformer "neutral" to ground is not feasible—as in the case of autotransformers. As described in (2), a trio of series blocking capacitors might have only 1% the cost of the series capacitors used for power-factor correction of long lines. The series blocking capacitors could be maintained shorted until potentially harmful GIC was detected, at which time the capacitors could be automatically and gracefully unshorted by silicon controlled rectifiers or other switches operating at the instant the voltage across the capacitor passes through zero.

**Can the market provide a more resilient bulk power system?**

FERC—the Federal Energy Regulatory Commission—and NERC—the North-American Electric Reliability Corporation—have a complex relationship themselves and with the organizations that generate, transmit, and distribute electric power in the United States and Canada. Thus far, the national interest in a more resilient bulk power system has not resulted in incentives or initiatives that would sufficiently advance that goal. The technical considerations discussed in this paper are important elements, but economic and organizational changes must be sought to result in the adoption of best world-wide practices in the North American Bulk Power System, and to advance beyond those best practices, where it is justified in the national interest.

**Testimony of Mrs. Bridgette Bourge**

**Senior Principal**

**National Rural Electric Cooperative Association**

**to the Committee on Homeland Security and Governmental Affairs**

**U.S. Senate**

**July 22, 2015**

## Introduction

Chairman Johnson, Ranking Member Carper, and members of the Committee, thank you for inviting me to testify today on "Protecting the Electric Grid from the Potential Threats of Solar Storms and Electromagnetic Pulse."

I serve as the lead legislative representative of the National Rural Electric Cooperative Association (NRECA) on homeland security issues. NRECA isthe service organization for over 900 not-for-profit electric utilities serving over 42 million people in 47 states. NRECA's members include 67 generation and transmission ("G&T") cooperatives that generate and transmit power to 668 of the 838 distribution cooperatives across the nation. Electric cooperative service territory makes up 75 percent of the nation's land mass. Kilowatt-hour sales by rural electric cooperatives account for approximately 11 percent of all electric energy sold in the United States. NRECA members generate approximately 50 percent of the electric energy they sell and purchase the remaining 50 percent.

As member owned not-for-profit utilities, distribution cooperatives and G&Ts reflect the values of their membership and they are uniquely focused on providing reliable electricity at the lowest reasonable cost. Cooperatives have to answer to their member/owners and justify every expense as they are the ones who will have to bear the cost. There is never any debate over whether a proposed project will benefit a cooperative's shareholders or customers because they are one and the same.

Today I am offering testimony on behalf of the electric industry to discuss two distinct issues: Geomagnetic Disturbances, or GMDs, and Electromagnetic Pulses, or EMPs.

## Clarifying the Terms

### Manmade EMPs

An EMP is a blast of electromagnetic energy that can disrupt or destroy electronic devices. There is a broad range of EMPs with significant variations in terms of impacts and responses. Just as the consequences and likelihood of each of these threats vary, so too does the approach to protecting the electric grid against them.

The only type of EMP that poses a potential widespread threat to the electrical grid are those generated by man through a high-altitude nuclear explosion. In the case of directed energy weapons or "suitcase EMPs" the threat is more localized, likely only impacting an individual facility similar to any other physical assault on grid infrastructure Due to the redundancy and resiliency of the grid, localized events like this have significantly less likelihood of causing a cascading electric system event.

The impact of an EMP from a high-altitude nuclear explosion over the United States would affect more than just electric infrastructure, however. Other critical infrastructures that utilize microprocessors are also vulnerable, including those with which the electric sector has interdependencies. Any activity that relies upon devices containing integrated circuitry- such as industrial process control systems, hospital equipment, as well as transportation and telecommunications systems - could be affected by an EMP attack on our country. As such, the primary responsibility for protecting the United States from such an attack should fall on the country's defense intelligence and military services, not on individual critical infrastructure owners/operators.

**Natural GMDs**

Geomagnetic disturbances caused by solar storms are initiated by natural events on the surface of the sun in which ejected masses of electrically charged particles are hurled toward the Earth. These create the potential for Earth-based disturbances due to their interaction with the Earth's magnetic field.

When the particles interact with the Earth's magnetic field, especially in certain geographic regions (e.g., northern latitudes), they can cause ground-induced currents (GIC) and other potentially disruptive phenomena. The direct impact of GMDs is primarily limited to reliability of the bulk power system and communication systems. GMDs are common and in fact happen pretty regularly. These are natural events and, as such, industry incorporates them into planning and mitigation efforts. Early alert systems using NOAA satellites allow owners and operators to take action to protect their systems, if necessary. With currently deployed satellites nearing the end of their reliable life cycle, these systems will need to be maintained and enhanced with new satellites in the near future to ensure that early alerts remain available and their timeliness is enhanced.

GMDs are ranked by storm levels, ranging from G1 (minor) to G5 (extreme). GMDs at higher levels have the potential to damage bulk power system assets (e.g., higher-voltage transformers) and to cause a loss of reactive power support, which could lead to voltage instability and power system collapse. The most significant issue for the bulk power system stemming from a strong GMD is the ability for operators to maintain voltage stability[1]. We see lower level GMDs pretty regularly. In fact, a few weeks ago we had a number of days in a row of G3 (strong) storms but no impact was felt on the bulk power system from these occurrences because proper measures were taken and procedures followed.

Based on these risks, in May 2013, under a statutory framework and authority established by Congress in the 2005 Energy Policy Act (Section 215 of the Federal Power Act), Federal Energy Regulatory Commission (FERC) Order No. 779 directed the North American Electric Reliability Corporation (NERC), to develop reliability standards to address the potential impact of GMDs on the reliable operation of the bulk electric system. NERC is an independent, government standards setting body that, under FERC oversight, develops and enforces mandatory reliability and critical infrastructure protection standards for the bulk power system owners, operators and users. NERC, FERC, and the electric power sector have since implemented a mandatory and enforceable GMD standard requiring owners and operators of the North American electric grid to prepare specific tailored operating procedures for use during severe GMD events. NERC has developed a second GMD standard, currently pending FERC approval, which will require tailored assessments and mitigation of the potential impacts of a 100-year GMD event on the

---

[1] Based upon the NERC 2012 Special Reliability Assessment Interim Report: Effects of Geomagnetic Disturbances on the Bulk Power System. http://www.nerc.com/files/2012GMD.pdf

bulk-power system, including high voltage power transformers. NRECA and its members support the approval and implementation plans developed in both of these standards

## Distinctions between EMPs and GMDs

Unfortunately, sometimes EMPs and GMDs are mistakenly conflated in the policy dialogue. It is important to keep these two separate threats distinct and to not conflate them from a policy, planning, protection or mitigation perspective. As stated earlier, EMPs are manmade and GMDs are caused by naturally occurring events. Furthermore, the magnetic fluctuations that result from GMDs are fundamentally different from EMPs generated by a high-altitude nuclear explosion and, as a result, pose different risks. Nuclear EMPs actually have three components—E1, E2, and E3—each of which arises from a different physical effect following a nuclear detonation. E3 is a slow pulse and resembles GMDs generated by a very severe solar flare. However, GMDs do not have an E1 or E2 component. The similarity between an EMP E3 component and a GMD caused by a sever solar flare may have led some to mistakenly confuse EMP and GMD, but such confusion overlooks critical distinctions and can have unintended consequences, including potentially undermining or conflicting with mitigation measures and protective standards already in place.

When considered as part of the broader spectrum of potential threats to the electric grid, nuclear-induced EMP is considered an extremely low-likelihood, high-consequence event. That doesn't mean the electric industry disregards or ignores its significance; merely that it is considered appropriately as part of a broader risk management strategy. The electric sector's approach to protecting critical assets against all types of threats is known as defense-in-depth, which includes balancing preparation, prevention, response, and recovery for a wide variety of hazards to electric grid operations. The industry recognizes that it cannot protect all assets from all threats. Instead, its priorities are to protect the most critical grid components against the most likely threats; to build in system resiliency; and to develop contingency plans for response and recovery when either man-made or natural phenomena impact grid operations.

Fundamentally, a nuclear-induced EMP would take the form of either a terrorist attack or an act of war occurring on or above U.S. soil. As such, the principal responsibility for preventing or guarding against a nuclear attack lies with the federal government. However, whatever the threat, industry works to ensure that the grid remains safe, and that reliable and affordable electricity is delivered to customers when and where they need it. We can't prevent every attack, remove every vulnerability, or respond in advance to every threat, but our defense-in-depth approach has proven successful in maintaining a highly reliable grid.

Industry works closely with government on matters of critical infrastructure protection through the Electric Sector Coordinating Council (ESCC). The ESCC brings together industry executives and senior-level government officials for high level policy discussions on important security issues affecting the electric industry. Both the public and private sector have unique roles, responsibilities, and capabilities. Leveraging each of these in a coordinated way is imperative. An EMP is the type of emerging issue that the ESCC can address at the policy level with DOE, DHS, the Department of Defense, and other federal agencies that have unique national resources beyond the capabilities of the private sector.

## Moving Forward

How do we minimize the potential consequences of an EMP or GMD?  Some propose that industry install their particular "protective device" or fully "gold plate" the entire grid so that it could, theoretically, at least partially survive a high altitude nuclear blast. However, there is no consensus on precisely what measures should be taken, the unintended effects they might have on the system, how much such an effort would cost, or how successful such efforts would be in limiting impacts to the bulk power system.  For example, due to non-uniform designs and complexity, substation solutions (e.g., Faraday-cages) would have to be individually customized, which would not come at a standardized rate. Additionally, there are concerns that installing "protective devices" in some areas of the bulk power system could unintentionally cause problems in other areas.  Further research and testing of these devices is needed, and is underway.

Even assuming that every conceivable blocking device were installed to protect every inch of the electric grid and caused no problems, power supplies still would likely be disrupted for a significant length of time in an impacted area.  That is because other critical infrastructures that utilities rely upon to function—such as transportation systems for our fuel, water systems for cooling, and telecommunications for operations—would also not be available.

The North American power grid is a huge, complex machine that spans the entirety of the United States, Canada and even parts of Mexico. Its function can be impacted by many different types of events or threats, from natural events like GMDs and severe storms to man-made malicious threats like EMP, cyber or physical attacks. Due to the expanse of not only these threats as well as the system itself, the electric sector addresses risk management through our defense-in-depth approach. This includes preparing for and preventing what we can, while at the same time planning for response and recovery in case of worst case scenarios.

Unfortunately, planning for recovery at a national level for widespread destructive events is necessary in today's world.  Efforts aimed at bolstering reserves of strategic transformers, for example, are a step in the right direction, as could be tasking DHS with further examination of EMP threats as a national security issue.

## Conclusion

Owners and operators of critical electric infrastructure have every incentive to prevent their systems from going down for even a moment if they can avoid it. Electric utility professionals know their systems best, including the operational and reliability impacts of potential external threats, so they should be included in any efforts commissioned to look into these matters. Utilizing existing public-private critical infrastructure partnership frameworks, like the Electricity Subsector Coordinating Council (ESCC), to ensure input and engagement on national security issues like recovery from a nuclear blast is, for a large part, why they exist.

Thank you for holding this hearing and inviting the electric industry to provide perspective on these very important issues and how they impact the complex machine that is the electric grid.  I would be happy to address any questions you may have.

www.ingramcontent.com/pod-product-compliance
Lightning Source LLC
Chambersburg PA
CBHW080611190526
45169CB00007B/2972